MACHINES

TASK CARD SERIES

Conceived and
written by
Ron Marson
Illustrated by
Peg Marson

10970 S. Mulino Rd.
Canby OR 97013

Dear Educator,
Please excuse our transition . . .

TOPS open-ended task card modules are taking on a new look. Task cards that used to come printed 4-up on heavy index card stock, packaged 2 sets to a zip-lock bag, are now printed 2-up at the back of this single book.

Even though our new cards are printed on lighter book stock, even though we haven't included an extra copy, we can now offer you something much better: You have our permission to make as many photocopies of these task cards as you like, as long as you restrict their use to the students you personally teach. This means you now can (1) incorporate task cards into full-sized worksheets, copying the card at the top of the paper and reserving the bottom for student responses. (2) You can copy and collate task card reference booklets, as many as you need for student use. Or (3) you can make laminated copies to display in your classroom, as before.

It will take some time to fully complete this transition. In the interim we will be shipping TOPS modules as a mixture of both old and new formats. Effective immediately (September 1989) this newer, more liberal photocopy permission applies to all task cards, including our older, heavier, 4-up standards!

Happy sciencing,

Ron Marson
author/publisher

Copyright © 1989 by TOPS Learning Systems. All rights reserved. Printed in the United States of America. No part of this book except the Reproducible Student Task Cards and Review/Test Questions may be duplicated, stored in a retrieval system, or transmitted, in any form or by any means, electronic, mechanical, photocopying, recording, or otherwise, without permission in writing from the publisher.

These Reproducible Student Task Cards may be duplicated for use with this module only, provided such reproductions bear copyright notice. Beyond single classroom use, reproduction of these task cards by schools or school systems for wider dissemination, or by anyone for commercial sale, is strictly prohibited.

ISBN 0-941008-99-1   Printed on Recycled Paper

CONTENTS

 **PART I** — **INTRODUCTION**

    A. A TOPS Model for Effective Science Teaching
    C. Getting Ready
    D. Gathering Materials
    E. Sequencing Task Cards
    F. Long Range Objectives
    G. Review / Test Questions

 **PART II** — **TEACHING NOTES**

    CORE CURRICULUM
1. Levers (1)
2. Work
3. Levers (2)
4. Paper Clip Pulley (1)
5. Paper Clip Pulley (2)
6. Wheel Pulley
7. Combination Pulley
8. Inclined Plane (1)
9. Inclined Plane (2)
10. What Kind of Machine?

    ENRICHMENT CURRICULUM
11. What Kind of Lever?
12. Super Pulley
13. Tug-of-War
14. Wheel and Axle
15. Spin Your Wheels!
16. Horsepower

 **PART III** — **REPRODUCIBLE STUDENT TASK CARDS**

Task Cards 1-16
Supplementary Pages — Protractor, Graph Paper

A TOPS Model for Effective Science Teaching...

If science were only a set of explanations and a collection of facts, you could teach it with blackboard and chalk. You could assign students to read chapters and answer the questions that followed. Good students would take notes, read the text, turn in assignments, then give you all this information back again on a final exam. Science is traditionally taught in this manner. Everybody learns the same body of information at the same time. Class togetherness is preserved.

But science is more than this.

Science is also process — a dynamic interaction of rational inquiry and creative play. Scientists probe, poke, handle, observe, question, think up theories, test ideas, jump to conclusions, make mistakes, revise, synthesize, communicate, disagree and discover. Students can understand science as process only if they are free to think and act like scientists, in a classroom that recognizes and honors individual differences.

Science is *both* a traditional body of knowledge *and* an individualized process of creative inquiry. Science as process cannot ignore tradition. We stand on the shoulders of those who have gone before. If each generation reinvents the wheel, there is no time to discover the stars. Nor can traditional science continue to evolve and redefine itself without process. Science without this cutting edge of discovery is a static, dead thing.

Here is a teaching model that combines the best of both elements into one integrated whole. It is only a model. Like any scientific theory, it must give way over time to new and better ideas. We challenge you to incorporate this TOPS model into your own teaching practice. Change it and make it better so it works for you.

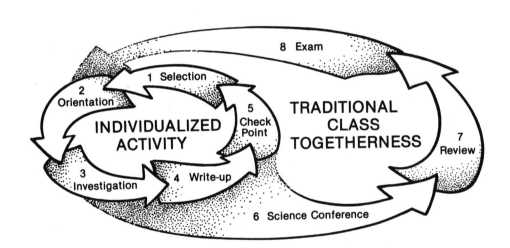

1. SELECTION

Doing TOPS is as easy as selecting the first task card and doing what it says, then the second, then the third, and so on. Working at their own pace, students fall into a natural routine that creates stability and order. They still have questions and problems, to be sure, but students know where they are and where they need to go.

Students generally select task cards in sequence because new concepts build on old ones in a specific order. There are, however, exceptions to this rule: students might *skip* a task that is not challenging; *repeat* a task with doubtful results; *add* a task of their own design to answer original "what would happen if" questions.

2. ORIENTATION

Many students will simply read a task card and immediately understand what to do. Others will require further verbal interpretation. Identify poor readers in your class. When they ask, "What does this mean?" they may be asking in reality, "Will you please read this card aloud?"

With such a diverse range of talent among students, how can you individualize activity and still hope to finish this module as a cohesive group? It's easy. By the time your most advanced students have completed all the task cards, including the enrichment series at the end, your slower students have at least completed the basic core curriculum. This core provides the common background so necessary for meaningful discussion, review and testing on a class basis.

3. INVESTIGATION

Students work through the task cards independently and cooperatively. They follow their own experimental strategies and help each other. You encourage this behavior by helping students only *after* they have tried to help themselves. As a resource person, you work to stay *out* of the center of attention, answering student questions rather than posing teacher questions.

When you need to speak to everyone at once, it is appropriate to interrupt individual task card activity and address the whole class, rather than repeat yourself over and over again. If you plan ahead, you'll find that most interruptions can fit into brief introductory remarks at the beginning of each new period.

4. WRITE-UP

Task cards ask students to explain the "how and why" of things. Write-ups are brief and to the point. Students may accelerate their pace through the task cards by writing these reports out of class.

Students may work alone or in cooperative lab groups. But each one must prepare an original write-up. These must be brought to the teacher for approval as soon as they are completed. Avoid dealing with too many write-ups near the end of the module, by enforcing this simple rule: each write-up must be approved *before* continuing on to the next task card.

5. CHECK POINT

The student and teacher evaluate each write-up together on a pass/no-pass basis. (Thus no time is wasted haggling over grades.) If the student has made reasonable effort consistent with individual ability, the write-up is checked off on a progress chart and included in the student's personal assignment folder or notebook kept on file in class.

Because the student is present when you evaluate, feedback is immediate and effective. A few seconds of this direct student-teacher interaction is surely more effective than 5 minutes worth of margin notes that students may or may not heed. Remember, you don't have to point out every error. Zero in on particulars. If reasonable effort has not been made, direct students to make specific improvements, and see you again for a follow-up check point.

A responsible lab assistant can double the amount of individual attention each student receives. If he or she is mature and respected by your students, have the assistant check the even-numbered write-ups while you check the odd ones. This will balance the work load and insure that all students receive equal treatment.

6. SCIENCE CONFERENCE

After individualized task card activity has ended, this is a time for students to come together, to discuss experimental results, to debate and draw conclusions. Slower students learn about the enrichment activities of faster students. Those who did original investigations, or made unusual discoveries, share this information with their peers, just like scientists at a real conference. This conference is open to films, newspaper articles and community speakers. It is a perfect time to consider the technological and social implications of the topic you are studying.

7. READ AND REVIEW

Does your school have an adopted science textbook? Do parts of your science syllabus still need to be covered? Now is the time to integrate other traditional science resources into your overall program. Your students already share a common background of hands-on lab work. With this shared base of experience, they can now read the text with greater understanding, think and problem-solve more successfully, communicate more effectively.

You might spend just a day on this step or an entire week. Finish with a review of key concepts in preparation for the final exam. Test questions in this module provide an excellent basis for discussion and study.

8. EXAM

Use any combination of the review/test questions, plus questions of your own, to determine how well students have mastered the concepts they've been learning. Those who finish your exam early might begin work on the first activity in the next new TOPS module.

Now that your class has completed a major TOPS learning cycle, it's time to start fresh with a brand new topic. Those who messed up and got behind don't need to stay there. Everyone begins the new topic on an equal footing. This frequent change of pace encourages your students to work hard, to enjoy what they learn, and thereby grow in scientific literacy.

GETTING READY

Here is a checklist of things to think about and preparations to make before your first lesson.

☐ Decide if this TOPS module is the best one to teach next.

TOPS modules are flexible. They can generally be scheduled in any order to meet your own class needs. Some lessons within certain modules, however, do require basic math skills or a knowledge of fundamental laboratory techniques. Review the task cards in this module now if you are not yet familiar with them. Decide whether you should teach any of these other TOPS modules first: *Measuring Length, Graphing, Metric Measure, Weighing* or *Electricity* (before *Magnetism*). It may be that your students already possess these requisite skills or that you can compensate with extra class discussion or special assistance.

☐ Number your task card masters in pencil.

The small number printed in the lower right corner of each task card shows its position within the overall series. If this ordering fits your schedule, copy each number into the blank parentheses directly above it at the top of the card. Be sure to use pencil rather than ink. You may decide to revise, upgrade or rearrange these task cards next time you teach this module. To do this, write your own better ideas on blank 4 x 6 index cards, and renumber them into the task card sequence wherever they fit best. In this manner, your curriculum will adapt and grow as you do.

☐ Copy your task card masters.

You have our permission to reproduce these task cards, for as long as you teach, with only 1 restriction: please limit the distribution of copies you make to the students you personally teach. Encourage other teachers who want to use this module to purchase their *own* copy. This supports TOPS financially, enabling us to continue publishing new TOPS modules for you. For a full list of task card options, please turn to the first task card masters numbered "cards 1-2."

☐ Collect needed materials.

Please see the opposite page.

☐ Organize a way to track completed assignment.

Keep write-ups on file in class. If you lack a vertical file, a box with a brick will serve. File folders or notebooks both make suitable assignment organizers. Students will feel a sense of accomplishment as they see their file folders grow heavy, or their notebooks fill up, with completed assignments. Easy reference and convenient review are assured, since all papers remain in one place.

Ask students to staple a sheet of numbered graph paper to the inside front cover of their file folder or notebook. Use this paper to track each student's progress through the module. Simply initial the corresponding task card number as students turn in each assignment.

☐ Review safety procedures.

Most TOPS experiments are safe even for small children. Certain lessons, however, require heat from a candle flame or Bunsen burner. Others require students to handle sharp objects like scissors, straight pins and razor blades. These task cards should not be attempted by immature students unless they are closely supervised. You might choose instead to turn these experiments into teacher demonstrations.

Unusual hazards are noted in the teaching notes and task cards where appropriate. But the curriculum cannot anticipate irresponsible behavior or negligence. It is ultimately the teacher's responsibility to see that common sense safety rules are followed at all times. Begin with these basic safety rules:

1. Eye Protection: Wear safety goggles when heating liquids or solids to high temperatures.
2. Poisons: Never taste anything unless told to do so.
3. Fire: Keep loose hair or clothing away from flames. Point test tubes which are heating away from your face and your neighbor's.
4. Glass Tubing: Don't force through stoppers. (The teacher should fit glass tubes to stoppers in advance, using a lubricant.)
5. Gas: Light the match first, before turning on the gas.

☐ Communicate your grading expectations.

Whatever your philosophy of grading, your students need to understand the standards you expect and how they will be assessed. Here is a grading scheme that counts individual effort, attitude and overall achievement. We think these 3 components deserve equal weight:

1. Pace (effort): Tally the number of check points you have initialed on the graph paper attached to each student's file folder or science notebook. Low ability students should be able to keep pace with gifted students, since write-ups are evaluated relative to individual performance standards. Students with absences or those who tend to work at a slow pace may (or may not) choose to overcome this disadvantage by assigning themselves more homework out of class.

2. Participation (attitude): This is a subjective grade assigned to reflect each student's attitude and class behavior. Active participators who work to capacity receive high marks. Inactive onlookers, who waste time in class and copy the results of others, receive low marks.

3. Exam (achievement): Task cards point toward generalizations that provide a base for hypothesizing and predicting. A final test over the entire module determines whether students understand relevant theory and can apply it in a predictive way.

Gathering Materials

Listed below is everything you'll need to teach this module. You already have many of these items. The rest are available from your supermarket, drugstore and hardware store. Laboratory supplies may be ordered through a science supply catalog.

Keep this classification key in mind as you review what's needed:

special in-a-box materials:	general on-the-shelf materials:
Italic type suggests that these materials are unusual. Keep these specialty items in a separate box. After you finish teaching this module, label the box for storage and put it away, ready to use again the next time you teach this module.	Normal type suggests that these materials are common. Keep these basics on shelves or in drawers that are readily accessible to your students. The next TOPS module you teach will likely utilize many of these same materials.
(substituted materials):	***optional materials:**
A parentheses following any item suggests a ready substitute. These alternatives may work just as well as the original, perhaps better. Don't be afraid to improvise, to make do with what you have.	An asterisk sets these items apart. They are nice to have, but you can easily live without them. They are probably not worth the an extra trip, unless you are gathering other materials as well.

Everything is listed in order of first use. Start gathering at the top of this list and work down. Ask students to bring recycled items from home. The teaching notes may occasionally suggest additional student activity under the heading "Extensions." Materials for these optional experiments are listed neither here nor in the teaching notes. Read the extension itself to find out what new materials, if any, are required.

Needed quantities depend on how many students you have, how you organize them into activity groups, and how you teach. Decide which of these 3 estimates best applies to you, then adjust quantities up or down as necessary:

$Q_1 / Q_2 / Q_3$

- **Single Student:** Enough for 1 student to do all the experiments.
- **Individualized Approach:** Enough for 30 students informally working in 10 lab groups, all self-paced.
- **Traditional Approach:** Enough for 30 students, organized into 10 lab groups, all doing the same lesson.

KEY:	*special in-a-box materials* (substituted materials)	general on-the-shelf materials *optional materials

$Q_1 / Q_2 / Q_3$

6/60/90	identical textbooks	1/1/1	*can opener
1/10/10	meter sticks	1/1/1	*nut cracker
1/10/10	large rubber stoppers	1/1/1	broom
1/1/1	gallon jug of water	1/10/10	feet of strong pliable wire, about 14 gauge
1/1/1	roll of masking tape	1/1/1	strong smooth post, fixed upright
1/10/10	rulers – centimeters or inches	20/20/20	feet of cord or rope
1/10/10	baby food jars with tight-fitting lids	1/10/10	medium-sized cans
1/1/1	roll of thread	4/40/40	rubber bands
1/10/10	*spring scales – 2 Newton capacity (200 grams) is ideal*	1/10/10	small lids from cooking oil bottles, or equivalent
1/1/1	roll of plastic wrap	1/10/10	large lids from mayonnaise jars, or equivalent
1/5/10	pairs of scissors	1/1/1	small nail or thumbtack
1/1/1	box of paper clips	1/1/1	hammer
1/10/10	*ring stands	2/20/20	straight pins
1/5/10	*single wheel pulleys*	1/1/1	roll of string
2/20/20	flexible plastic drinking straws	1/1/1	flight of stairs
1/1/2	*paper punches	1/1/1	yard stick (foot ruler and string)
1/10/10	*pieces of cardboard cut from boxes – about as long and wide as notebook paper*	1/1/1	stopwatch
2/20/20	index cards – 4x6 inch work best	1/1/1	*bathroom scale*
1/1/2	staplers	1/5/10	*hand calculators
1/1/1	*bolts		

D

Sequencing Task Cards

This logic tree shows how all the task cards in this module tie together. In general, students begin at the trunk of the tree and work up through the related branches. As the diagram suggests, the way to upper level activities leads up from lower level activities.

At the teacher's discretion, certain activities can be omitted or sequences changed to meet specific class needs. The only activities that must be completed in sequence are indicated by leaves that open *vertically* into the ones above them. In these cases the lower activity is a prerequisite to the upper.

When possible, students should complete the task cards in the same sequence as numbered. If time is short, however, or certain students need to catch up, you can use the logic tree to identify concept-related *horizontal* activities. Some of these might be omitted since they serve only to reinforce learned concepts rather than introduce new ones.

On the other hand, if students complete all the activities at a certain horizontal concept level, then experience difficulty at the next higher level, you might go back down the logic tree to have students repeat specific key activities for greater reinforcement.

For whatever reason, when you wish to make sequence changes, you'll find this logic tree a valuable reference. Parentheses in the upper right corner of each task card allow you total flexibility. They are left blank so you can pencil in sequence numbers of your own choosing.

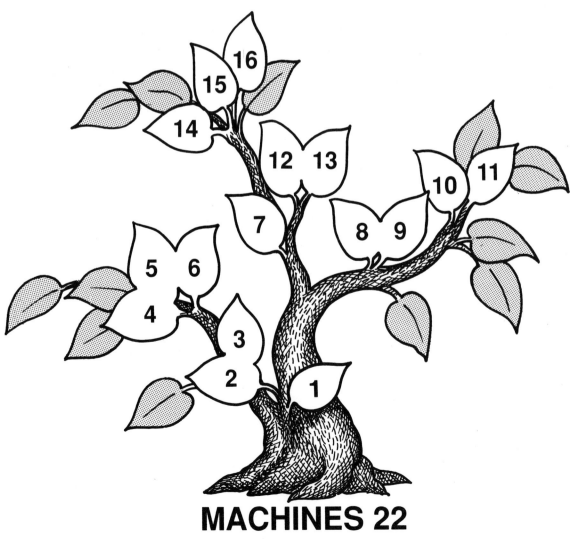

MACHINES 22

E

LONG-RANGE OBJECTIVES

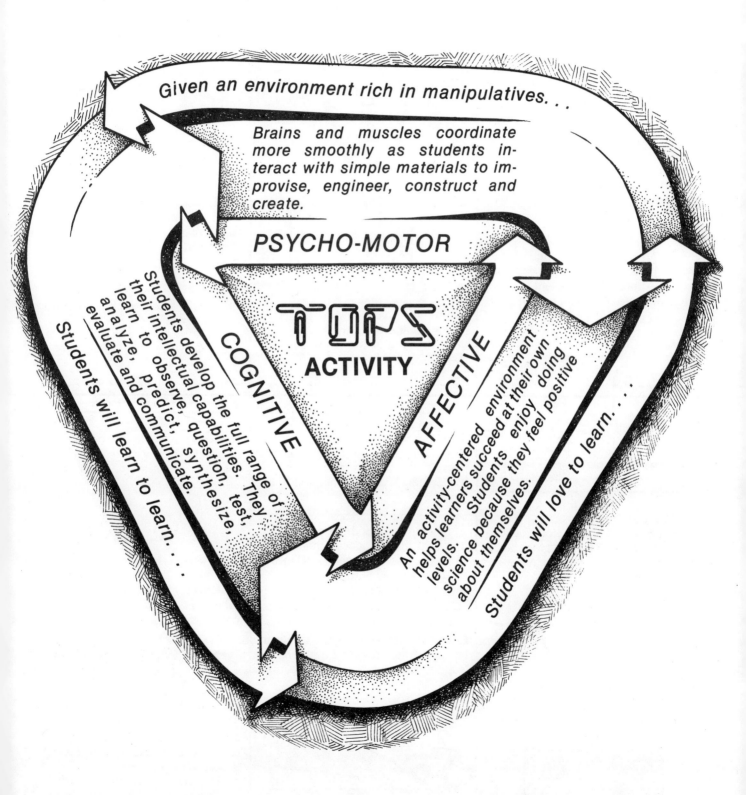

Review / Test Questions

Photocopy the questions below. On a separate sheet of blank paper, cut and paste those boxes you want to use as test questions. Include questions of your own design, as well. Crowd all these questions onto a single page for students to answer on another paper, or leave space for student responses after each question, as you wish. Duplicate a class set and your custom-made test is ready to use. Use leftover questions as a review in preparation for the final exam.

task 1
Draw how to move a large rock using a strong wood pole and a brick. Label the effort, fulcrum and resistance.

task 2
Which is more work — lifting 50 lbs of sand 6 feet or 100 lbs of sand 3 feet?

task 3
A 100 lb girl and a 150 lb boy balance each other on a see-saw. The girl raises the boy 2 feet higher.
a. Calculate how much work she did.
b. Assuming the lever is 100% efficient, how far did the girl move down? Show your math.

tasks 4-5
A 20 Newton weight is attached to each pulley system. Assuming both systems are 100% efficient…
a. How hard must you pull on each rope to lift the resistance?
b. How much work lifts each resistance 3 meters?

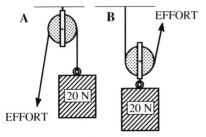

tasks 5-6
In this single movable pulley system, the rope is pulled with a force of 30 lb through 4 feet. The 45 lb resistance is raised 2 feet. Calculate the…
a. work input.
b. work output.
c. efficiency of the pulley.

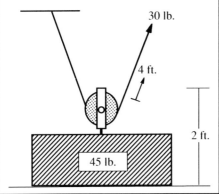

tasks 5-7
In this combination pulley system, the rope moves 4 times as far as the resistance. How much effort is required to lift a 60 N resistance if the machine is…
a. friction free?
b. 25% efficient?

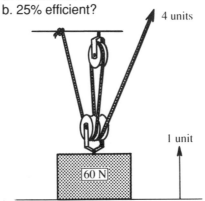

task 6
An engineer is asked to design a single movable pulley that is fully 100% efficient. Why is this an impossible task? Give 2 reasons.

tasks 8-9
An inclined plane measures 12 meters long and 2 meters high.
a. Assuming no friction, what effort is required to push a 420 N cart up this plane?
b. How much work is required to do this?

task 3-9
It is said that "machines make less work." Is this statement true in the scientific sense? Support your answer with an example.

tasks 4-9
Can this man pull his 2000 lb car up an incline 20 feet long and 2 feet high? Support your answer with calculations.

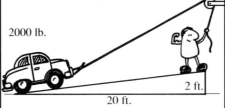

task 10
Classify each machine as a lever or inclined plane. Give reasons for your answer.
a. A wheelbarrow.
b. A wedge.
c. A pair of pliers.
d. A steering wheel.
e. A screw

task 11
What class of lever is a fly swatter? Defend your answer.

task 12-13
How much rope must this worker pull through the pulley to raise the resistance 1 meter? Explain your reasoning.

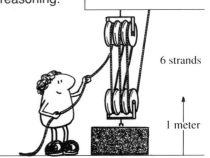

task 14
A bicycle has pedals 20 cm in diameter connected to a sprocket 4 cm in diameter. Calculate the ideal mechanical advantage of this machine.

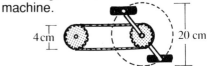

task 15
Which bicycle is easier to pedal? Which bicycle goes faster? Defend your answer.

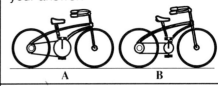

task 16
Jack and Jill both climb a hill. Jack runs and Jill walks. If Jack's weight equals Jill's…
a. Who did the most work? Explain.
b. Who used the most power? Explain.

Copyright © 1989 by TOPS Learning Systems.

Answers

task 1

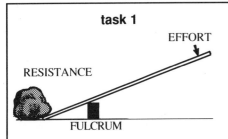

task 2
The amount of work is the same in each case.
Work = force x distance
= 50 lbs x 6 feet
= 100 lbs x 3 feet
= 300 foot-lbs

task 3
a. Work done = 150 lbs x 2 feet
= 300 ft-lbs
b. Assuming 100% efficiency…
$Work_{in} = Work_{out}$
100 lbs x distance = 300 ft-lbs
distance = 3 feet

task 4-5
a. A force of 20 N applied to rope **A** lifts its resistance. A force of 10 N applied to rope **B** lifts its resistance.

b. $Work_{out}$ = force x distance
= 20 N x 3 m = 60 N-m.
This much work is required by either system.

task 5-6
a. $work_{in}$ = force x distance
= 30 lbs x 4 feet = 120 ft-lbs
b. $work_{out}$ = force x distance
= 45 lbs x 2 feet = 90 ft-lbs
c. efficiency = $\frac{work_{out}}{work_{in}}$ x 100
= $\frac{90 \text{ ft-lbs}}{120 \text{ ft-lbs}}$ = 75 %

task 5-7
a. Without friction, $work_{in} = work_{out}$.
effort x 4 units = 60 N x 1 unit
effort = 15 N

b. With a 100 % friction-free pulley, 15 N of effort is required to lift the load. But since the pulley is only 25% efficient, 4 times as much effort, or 60 N, is required.

task 6
The weight of a moveable pulley that is lifted with its load, and the friction of its moving parts, both reduce its overall efficiency. An engineer can design a lightweight pulley, but not one that weighs nothing. Friction can be reduced, but not eliminated completely.

task 8-9
a. Without friction, $work_{in} = work_{out}$.
force x 12 meters = 420 N x 2 meters
force = 70 N
b. $work_{in} = work_{out}$
= 420 N x 2 meters
= 840 N-m

task 3-9
In a scientific sense, the statement is false. If the machine is less than 100% efficient, you have to put more work in than you get out. Consider this lever as an example.

Assuming 100% efficiency, 10 lbs of force lifts a 100 lb load 1 foot off the ground. Though the effort required to lift the load is reduced to 1/10, the distance through which you must apply this reduced effort increases 10 times. Hence, the amount of actual work accomplished remains constant.

$work_{in} = work_{out}$
10 lbs x 10 feet = 100 lbs x 1 foot

In practice, machines are less than 100% efficient, so that something more than 10 lbs would be necessary to lift the load, say 11 lbs. Even though this effort is still much less than the load, the work you put into the lever is now greater than the work you get out.

$work_{in}$ = 11 lbs x 10 ft > 100 lbs x 1 ft

task 4-9
Assuming 100% efficiency,
$work_{in} = work_{out}$
force x 20 feet = 2000 lbs x 2 feet
force = 200 lbs

A man heavier than 200 lbs can pull the car by applying all his weight to the rope. A lighter man cannot.

task 10
a. The wheelbarrow is a lever. Its wheel axle is the fulcrum; its load is the resistance; effort is applied at the handles.
b. The wedge is a plane inclined very slightly on both sides.
c. The pair of pliers is 2 levers joined at the fulcrum.
d. The steering wheel is a lever. Effort is applied at the outside of the wheel to the resistance at the wheel's shaft. The fulcrum is at the center.
e. The screw is a spiraling inclined plane.

task 11
The fly swatter is a 3rd class lever. The effort you apply at the handle comes between the pivot at your wrist and the resistance at the swatter end. This gives a distance advantage characteristic of 3rd class levers, moving the swatter fast enough to catch the fly, sometimes.

task 12-13
The resistance is supported by 6 strands of rope. Thus the rope must be pulled 6 times farther than the resistance is lifted, or 6 meters.

Task 14
$IMA = \frac{\text{Distance Effort Moves}}{\text{Distance Resistance Moves}}$
$= \frac{C}{c} = \frac{\pi D}{\pi d} = \frac{20 \text{ cm}}{4 \text{ cm}} = 5$

task 15
Bicycle **B** is easier to pedal. Its back wheel sprocket and pedal wheel have about the same diameter, so that 1 turn of the pedal results in one turn of the back wheel. Bicycle **A** goes faster. Its back wheel sprocket is much smaller than its pedal wheel, so that 1 turn of the pedal results in many turns of the back wheel.

task 16
a. Jack and Jill both raised the same amount of weight (their bodies) an equal distance. Hence each did the same amount of work.
b. Jack used more power than Jill because he did his equal amount of work in much less time. He was breathing much harder to reach the top of the hill first.

TEACHING NOTES
For Activities 1-16

Task Objective (TO) build a simple lever. To experience how this lever either reduces effort, or reduces the distance through which the effort is applied.

LEVERS (1)   ○   **Machines ( )**

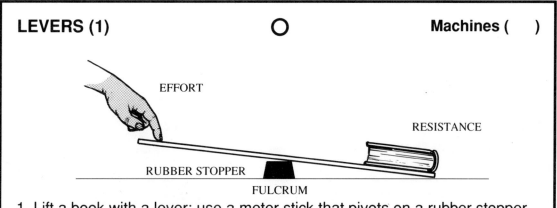

1. Lift a book with a lever: use a meter stick that pivots on a rubber stopper.

    a. Tell how to lift the resistance (the book) using a smaller effort (a smaller push from your hand). Draw a diagram.
    b. Tell how to lift the same resistance using a larger effort. Draw a diagram.

2. Did you apply the smaller effort through the same distance as the larger effort to lift the book? Explain.

© 1989 by TOPS Learning Systems

Answers / Notes

1a. Slide the fulcrum nearer the resistance to lift it with a smaller effort.

1b. Slide the fulcrum nearer the effort to lift the resistance with a larger effort.

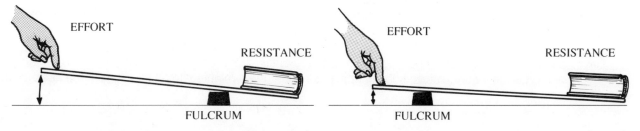

2. No. The smaller effort was applied through a relatively longer distance than the larger effort, as represented by the two arrows above.

Materials

☐ A meter stick.
☐ A large rubber stopper.
☐ A book.

(**TO**) understand the idea of "work" in a scientific sense.

WORK Machines ( )

1. To do "work" in the scientific sense, you must apply a *force* through a *distance*. In each of the following activities decide if you are doing work. Explain your reasoning.
   a. Lift a book off the table.
   b. Hold a book perfectly still at arm's length for about 1 minute.
   c. Push a book across the table.
   d. Push against a wall as hard as you can.

WORK = Force x Distance

2. Call the force necessary to lift 1 book a "bk." Call the distance of your extended hand 1 "span."
   a. How much work is required to lift 2 books up 3 spans?
   b. How much work is required to lift 6 books up 1 span?
   c. Which requires more work, lifting 3 books up 3 spans or 2 books up 4 spans? Explain.

© 1989 by TOPS Learning Systems

Introduction

Ask a small female student to lift a gallon jug of water (about 8 pounds) from the floor to a height of 6 feet. Calculate the work that she did on your blackboard:

W = force x distance = 8 lbs x 6 ft = 48 ft-lbs.

Now ask a large male student to hold the same jug of water at arms length for several minutes, as steady as he can. Calculate the work that he did on your blackboard:

W = force x distance = 8 lbs x 0 ft = 0 ft-lbs.

Develop the scientific idea of "work." Discuss how the scientific definition of work is different from common usage. (To use equivalent metric units, substitute a 4 liter water jug (weighing 36 N) lifted through a distance of 2 meters.)

Answers / Notes

1a. Work is done. A force is exerted on the book that lifts it through a distance.
1b. No work is done. The book does not move.
1c. Work is done. A force is exerted on the book that slides it through a distance.
1d. No work is done. The wall does not move.
2a. Work = force x distance = 2 bk x 3 spans = 6 bk-spans
2b. Work = force x distance = 6 bk x 1 spans = 6 bk-spans
2c. More work is required to lift 3 books a distance of 3 spans (9 bk-spans) than to lift 2 books a distance of 4 spans (8 bk-spans).

Materials

☐ A gallon jug of water.
☐ 6 identical textbooks.

(TO) recognize that a lever reduces the effort required to lift a resistance but does not reduce the work.

LEVERS (2)   Machines ( )

1. Use masking tape to fix a rubber stopper to the 50 cm center of a meter stick and label it "C". Tape and label the left 10 cm mark "L" and the right 70 cm mark "R."

2. Balance some books against each other at points L and R. What can you discover?

3. Measure how far each point (L and R) move up and down. What can you discover?

4. If you push down with a force of 3 books at L…
   a. How many books can you just begin to lift up at R?
   b. Does this lever reduce your effort? Explain.
5. Suppose you apply this 3 book force down through a distance of .2 hand spans… a. Calculate the work you put *in* at L and the work you get *out* at R.
   b. Did the lever reduce your work? Explain.

© 1989 by TOPS Learning Systems

Answers / Notes

2. One book at L balances 2 books at R. Two books at L balance 4 books at R. In general, any load at L balances twice the load at R.

3. Point L moves about 4 cm, while point R moves about 2 cm. In general, L moves though twice the distance of R.

4a. A 3 bk force applied down will just begin to lift up 6 books (twice as many).
4b. Yes, the lever reduces your effort, since a 3 bk force lifts a 6 bk load.

5a. $Work_{in}$ = 3 bks x .2 spans = .6 bk-spans
    $Work_{out}$ = 6 bks x .1 span = .6 bk-spans
5b. No, $W_{in} = W_{out}$. Half the effort was applied, but through twice the distance.

Materials
☐ A meter stick.
☐ A large rubber stopper.
☐ Masking tape.
☐ 9 identical textbooks.
☐ A ruler.

(TO) construct a weight for measuring on a spring scale. To compare a movable pulley with a fixed pulley.

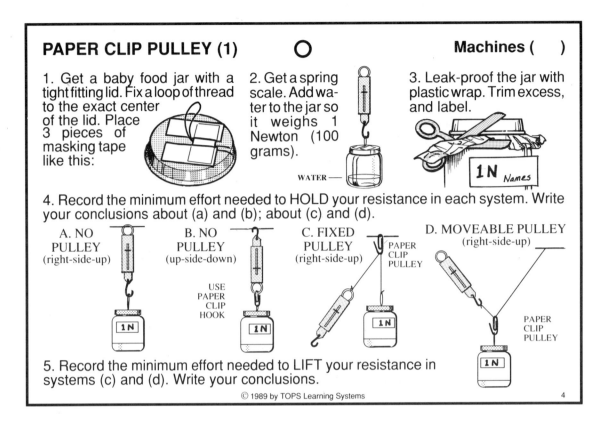

Introduction
(Please see the introductory notes printed under task 5 to the right.)

Answers / Notes
1. *Position the tape so its edge remains on center. Notice how it is reinforced by 2 other pieces of tape.*

2. *Students will be lifting the jar by its lid. Caution them to keep it near the table surface, in case it falls.*

3. *Label the jar with its weight and the names of lab group members. If you have different kinds of spring scales, label them as well, so the same scale will be used with the same weight throughout this module.*

4. (*These results are based on a load of 1 Newton. If your scales are calibrated in earth-equivalent grams, multiply all answers here, and elsewhere, by 100.*)

Systems (A) and (B) read 1.0 N and 1.2 N respectively. This shows that the spring scale reads too high when used up-side-down, adding the weight of the scale casing to the weight of the load.

Systems (C) and (D) read .15 N and .35 N respectively. Friction supports much of the 1 N load in each pulley, more in the fixed pulley than in the movable pulley.

5. Systems (C) and (D) read 1.15 N and .60 N respectively. A movable pulley allows you to lift the load with about half as much effort as the fixed pulley. In each case, a little extra force is required to overcome friction.

Materials
☐ A baby food jar with tight fitting lid.
☐ Thread.
☐ Masking tape.
☐ A spring scale. 2 Newton capacity (200 grams) is ideal.
☐ Water.
☐ Plastic wrap.
☐ Scissors.
☐ Paper clips.
☐ A ring stand is optional. A table edge also serves as a fixed support, or pulley systems can be hand held.

(TO) recognize that a simple movable pulley reduces the effort required to lift a resistance, but does not reduce the work.

PAPER CLIP PULLEY (2)   Machines ( )

1. How much work do you put *in* the moveable paper clip pulley when you pull the string a distance of 2 hand spans?

2. How much work does this pulley give *out* as it lifts the resistance?

3. Calculate the *efficiency* of the pulley.

$$\% \text{ Efficiency} = \frac{\text{Work}_{out}}{\text{Work}_{in}} \times 100$$

4. Does the pulley give…
   a. an effort advantage? Explain.
   b. a work advantage? Explain.

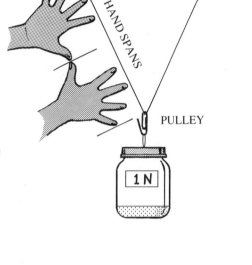

© 1989 by TOPS Learning Systems

Introduction

(These introductory notes pertain to task 4 to the left. They are placed on this side of the page to save space.)

Make a baby-food-jar weight in advance. Demonstrate how to add water and reweigh until reaching the correct weight. Scales with 1.0 Newton capacities (≈100 grams), on up to heavier ones with 2.5 Newton capacities (≈250 grams), can all be conveniently paired with 1 Newton jars. If your spring scales are heavier, fill the jars with more water, perhaps 2.0 Newtons (≈200 grams). You may substitute other weights besides baby food jars. However, they must be cylinderical in order to roll on inclined planes in tasks 8 and 9.

Now is the time to explain the difference between weight and mass. Weight, arising from the force due to gravity, is correctly measured in Newtons (the force required to accelerate 1 kilogram of mass 1 meter every second). Mass, the quantity of matter in a body, is correctly measured in grams (the matter in 1 cu cm of water). On earth, 100 grams of matter weighs about 1 Newton (.98 Newtons to be more exact). But in space, this same 100 grams mass weighs 0 Newtons, nothing at all. Spring scales that are calibrated in grams instead of Newtons should be understood as earth-bound. They are invalid on the moon, Jupiter and many other places.

Answers / Notes

(Answers may vary somewhat, depending on frictional forces that are generated as your particular brand of thread slides around its paper clip.)

1. A .60 N effort is required to lift the 1.00 N resistance.
   Work_{in} = .60 N x 2 spans = 1.20 N-spans

2. The resistance moves only one span as the effort is applied through 2 spans.
   Work_{out} = 1.00 N x 1 span = 1.00 N-spans

3. $\% \text{ Efficiency} = \frac{\text{Work}_{out}}{\text{Work}_{in}} \times 100 = \frac{1.00 \text{ N-spans}}{1.20 \text{ N-spans}} \times 100 = 83\%$

4a. Yes. The 1.00 N resistance was raised by only a .60 N effort.
4b. No. While 1.20 N-spans of work was put into the pulley, it only did 1.00 N-spans of work on the resistance. The difference, .20 N-spans, was lost to frictional drag as the thread slid around the paper clip.

Materials

☐ Materials from task card 4.

(TO) evaluate the efficiency of a movable wheel pulley. To recognize that work output is reduced by the weight of the pulley and frictional drag.

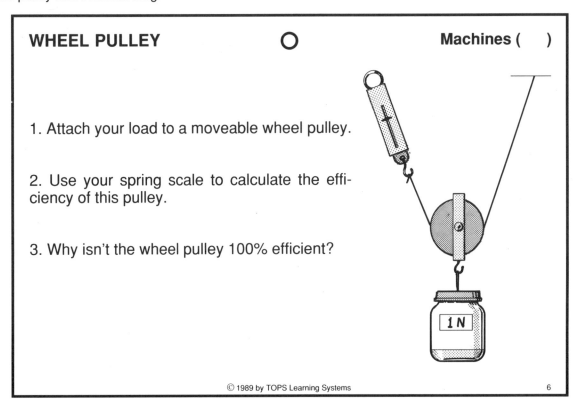

WHEEL PULLEY ○ Machines ( )

1. Attach your load to a moveable wheel pulley.

2. Use your spring scale to calculate the efficiency of this pulley.

3. Why isn't the wheel pulley 100% efficient?

© 1989 by TOPS Learning Systems

Answers / Notes

2. Students should first apply an effort through any measured distance (hand spans, for example), then repeat the series of calculations they did in task 5.

3. The pulley is less than 100% efficient because extra work is required to lift the weight of the pulley, as well as the load. Frictional drag is a less important, almost negligible factor. *(Some students may count the extra weight of the pulley as part of the total resistance. In this case the efficiency of a freely turning pulley will be nearly 100%.)*

Materials
☐ A wheel pulley.
☐ Materials from task card 4.

(TO) evaluate the efficiency of a combination paper clip pulley. To calculate the effort required to lift the resistance under ideal frictionless conditions.

COMBINATION PULLEY ○ Machines ( )

1. Twist two paper clips apart as shown.

2. Build a double paper clip pulley.

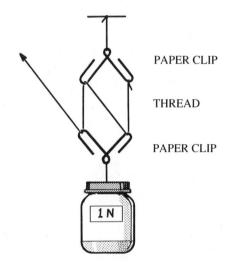

PAPER CLIP
THREAD
PAPER CLIP
1 N

3. Calculate the efficiency of this machine. Show your work.

4. Suppose this pulley were friction free so that Work$_{in}$ = Work$_{out}$. What effort would be needed to lift your load? Explain your reasoning.

© 1989 by TOPS Learning Systems   7

Answers / Notes

3. *(Answers will vary, depending on the friction that results as the thread slides around the paper clips. Here is one result.)*

A .50 N effort is required to lift the 1.00 N resistance. The effort moves through a distance of 4 spans to lift the resistance a distance of 1 span.

$$\text{Work}_{in} = .50 \text{ N} \times 4 \text{ spans} = 2.00 \text{ N-spans}$$
$$\text{Work}_{out} = 1.00 \text{ N} \times 1 \text{ span} = 1.00 \text{ N-spans}$$
$$\% \text{ Efficiency} = \frac{\text{Work}_{out}}{\text{Work}_{in}} \times 100 = \frac{1.00 \text{ N-span}}{2.00 \text{ N-span}} \times 100 = 50\%$$

4. If work$_{in}$ = work$_{out}$...

$$\text{force}_{in} \times \text{distance} = \text{force}_{out} \times \text{distance}$$
$$\text{force}_{in} \times 4 \text{ spans} = 1.0 \text{ N} \times 1 \text{ span}$$
$$\text{force}_{in} = .25 \text{ N}$$

Materials
☐ Materials from task card 4.

(TO) graph how the effort required to pull a cart up an inclined plane changes with its angle of inclination.

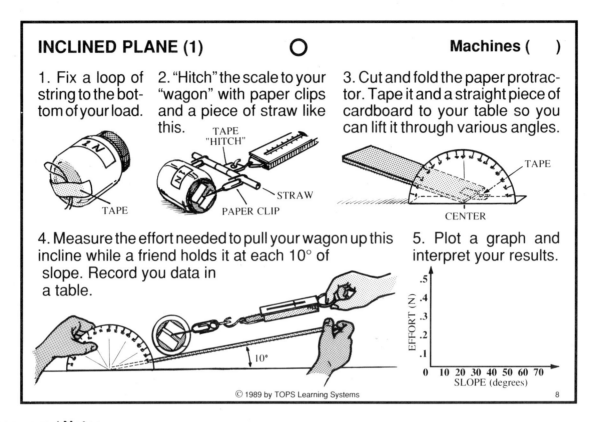

Answers / Notes

2. *Keep the 2 paper clips well separated on the straw to prevent them from dragging over the top and bottom surfaces of the jar as it rolls up the ramp.*

3. *The top surface of the incline shows through the paper protractor as a distinct shadow. To properly align the protractor's image to this shadow, place its center slightly beyond the end of the ramp, where the incline's surface and the table's surface intersect.*

4-5. *The effort required to pull a resistance up an incline increases with increasing slope. This increase is less rapid as the angle of inclination approaches 90°. (This effort, in quantitative terms, is given by w sin ø, where w is the weight of the wagon and ø is the angle of inclination.)*

In practice, it is difficult to read a moving scale with any great accuracy. Moreover, it may actually read too light at low ramp angles. Holding the scale in a near horizontal position tends to increase its internal friction as moving parts slide over fixed parts. This graph line maps the theoretical sine curve. Notice how actual data points scatter significantly from this ideal.

Materials

☐ The baby-food-jar weight constructed in task card 4.
☐ Masking tape, thread, and paper clips.
☐ A plastic straw.
☐ A paper punch (optional) to use in step 2. Or drill a hole through the hitch with a pin and pencil.
☐ Scissors.
☐ A flat piece of cardboard.
☐ A spring balance.
☐ Graph paper and a protractor. Both may be photocopied from the back of this book.

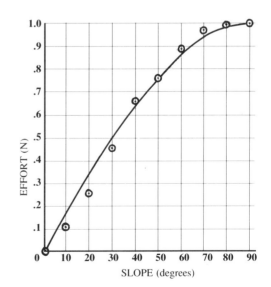

(TO) recognize that an inclined plane reduces the effort required to lift a resistance, but does not reduce the work.

INCLINED PLANE (2) O Machines ( )

1. Divide two 4x6 index cards exactly in half the long way. Make a T-square from 3 pieces, and a 3-unit ruler from the last piece.

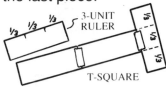

2. Use your tools to draw these 4 right triangles on scratch paper, then cut them out.

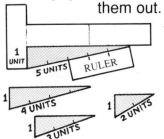

3. Use your protractor to fill in the value of all angles…

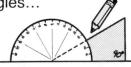

…Staple your triangles together like these.

4. Turn the triangles over to form 4 inclines for angling your cardboard ramp.

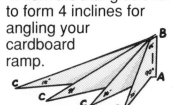

a. How much force must you use to lift your load from level A through 1 unit to level B? How much work does this require?

b. Assume this work is the same for each incline. Calculate the force you need to move your load from points C up 2, 3, 4, and 5 units to level B.

c. Check your answer on the 30° ramp.

© 1989 by TOPS Learning Systems

Answers / Notes

1. *4x6 index cards divided exactly in half yield 2x6 strips. Three strips placed side by side define the ruler divisions, and accurate placement of the T's leg. This won't work, of course, for cards cut to other proportions.*

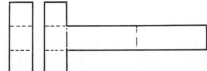

3. *Students unaware that the internal angles of a triangle add to 180° should be asked to add them up. This is a good way to verify the accuracy of each measured angle.*

5 unit triangle: 90.0° + 78.5° + 11.5° = 180.0°
4 unit triangle: 90.0° + 75.5° + 14.5° = 180.0°
3 unit triangle: 90.0° + 70.5° + 19.5° = 180.0°
2 unit triangle: 90.0° + 60.0° + 30.0° = 180.0°

4a. The load weighs 1.00 N. To lift it through a distance of 1 unit…
    Work = force x distance = 1.00 N x 1 unit = 1 N-unit

4b. force x 5 units = 1 N-unit   force x 3 units = 1 N-unit
         force = 1/5 N = .20 N   force = 1/3 N = .33 N
     force x 4 units = 1 N-unit   force x 2 units = 1 N-unit
         force = 1/4 N = .25 N   force = 1/2 N = .50 N

4c. The 30° slope theoretically cuts the required effort to half the load weight. Actual measurements may be 10 to 20% less (perhaps .40 N), because internal friction in many scales increases as they are tipped horizontally.

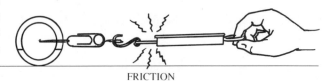

FRICTION

Materials

☐ 4x6 index cards. You may substitute other sizes if width to length have the same 1 to 3 proportion.
☐ Scissors and a stapler.
☐ Materials from task card 8.

(TO) classify common machines as levers or inclined planes.

WHAT KIND OF MACHINE?   O   Machines ( )

1. Classify each machine as a *lever* or *inclined plane*. Give reasons for your answer.

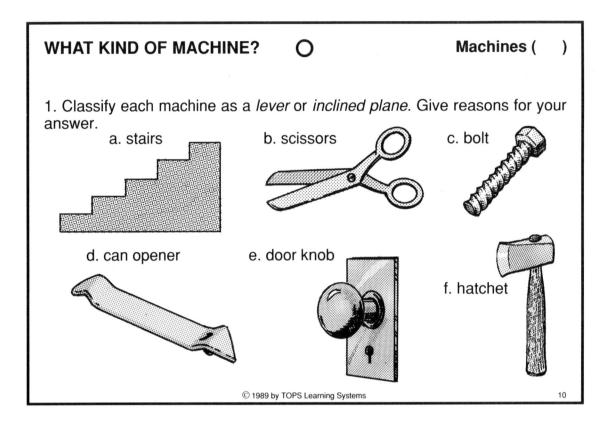

a. stairs   b. scissors   c. bolt
d. can opener   e. door knob   f. hatchet

© 1989 by TOPS Learning Systems

Answers / Notes

a. Inclined plane: A flight of stairs is a series of steps cut into an inclined plane.

b. Lever: A pair of scissors consists of 2 levers pivoting about a central fulcrum.

c. Inclined plane: A bolt is an inclined plane wound upon a cylinder.

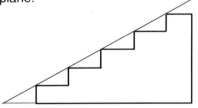

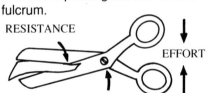

d. Lever: A can opener is a lever that pivots about the rim of a can.

e. Lever (wheel and axle): Applying effort to the outside of a door knob moves resistance in the shaft about a central pivot.

f. Inclined plane and lever: The blade is a double inclined plane (a wedge). Swung by the handle it becomes a lever.

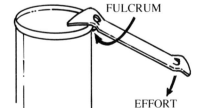

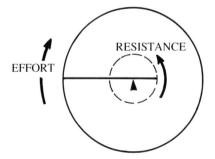

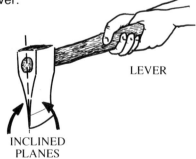

Materials
☐ Scissors, a bolt and a can opener (optional).

notes 10

(TO) study three different classes of levers. To classify common machines as to lever type.

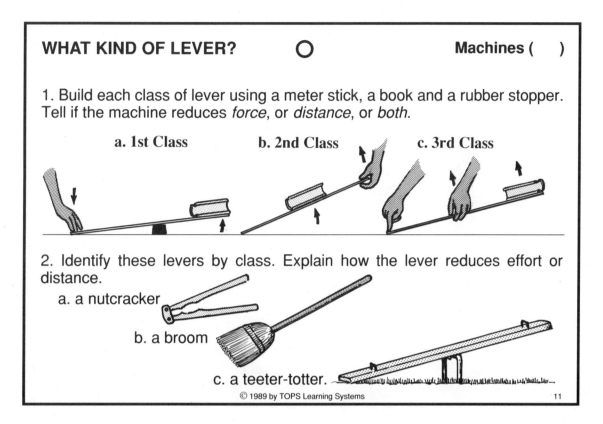

Answers / Notes

1a. The first class lever reduces effort or distance. If the effort arm is longer than the resistance arm, then a small effort can lift a heavy resistance. If the effort arm is shorter than the resistance arm, then a large effort applied through a short distance can lift a light resistance through a long distance.

1b. The 2nd class lever reduces effort. A small effort applied over a long distance lifts a heavy resistance through a short distance.

1c. The 3rd class lever reduces distance. A large effort applied over a short distance lifts a light resistance through a long distance.

2a. A nut cracker is a 2nd class lever that reduces effort. A small force applied at the end of the lever overcomes a large resistance located between the effort and the fulcrum.

2b. A broom is a 3rd class lever that reduces distance. As you sweep the floor, one hand holds the top of the broom handle stationary, acting as a pivot. The lower hand moves through a small distance to sweep the head of the broom through a longer distance.

2c. A teeter-totter is a 1st class lever that can reduce effort or distance (but not both at the same time).

  A small child who sits far from the fulcrum can lift a heavier adult who sits near the fulcrum. The child has an effort advantage.

  Moving slightly back, the adult can move down through a small distance to lift the smaller child up through a longer distance. The adult has a distance advantage.

Materials
☐ A meter stick.
☐ A rubber stopper.
☐ A book.
☐ A nutcracker and broom (optional).

(TO) relate the number of strands that support a pulley to the relative distances moved by the effort and resistance.

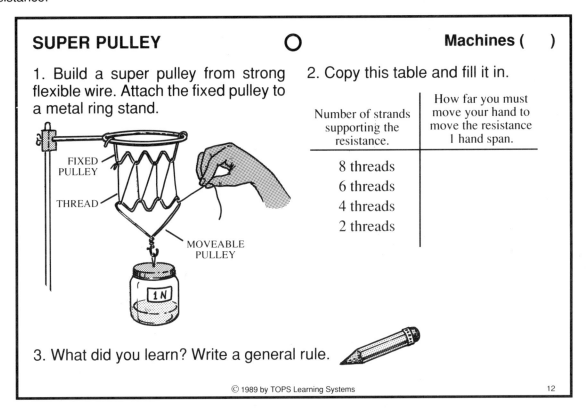

SUPER PULLEY — Machines ( )

1. Build a super pulley from strong flexible wire. Attach the fixed pulley to a metal ring stand.

2. Copy this table and fill it in.

Number of strands supporting the resistance.	How far you must move your hand to move the resistance 1 hand span.
8 threads	
6 threads	
4 threads	
2 threads	

3. What did you learn? Write a general rule.

© 1989 by TOPS Learning Systems

Answers / Notes

2.

Number of strands supporting the resistance.	How far you must move your hand to move the resistance 1 hand span.
8 threads	8 spans
6 threads	6 spans
4 threads	4 spans
2 threads	2 spans

3. Given a pulley with "n" support strands, the applied effort moves "n" times further than the resistance. *(The number "n" must always be even. The last odd string never supports the load directly. It only changes the direction of the applied force.)*

In other words, a pulley with "n" support strands has an ideal mechanical advantage of "n." Mechanical advantage will be introduced in task card 14.

Materials

☐ Strong flexible wire, about 14 gauge.
☐ A ring stand with metal ring attachment. Or securely tape the top half of the super pulley to the edge of a table like this:
☐ Thread.
☐ A paper clip.
☐ The baby-food-jar resistance.

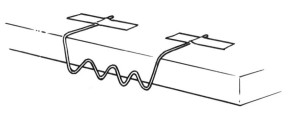

(TO) have a tug-of-war contest using a pulley system. To analyze force and distance advantages.

TUG-OF-WAR　　　　　　　　　　　　　　　　　**Machines ( )**

1. Locate a strong, smooth post. You may need to go outside to find one.

2. Wind a rope around the post and a broom handle as illustrated.

3. Have a tug-of-war. You grip the broom at (a) and (b). Have a friend pull the rope at (c).
    a. Who wins the tug-of-war?
    b. Who moves through the shortest distance?

4. Compare this rope system to the super pulley you made before.

© 1989 by TOPS Learning Systems　　　　　　　　13

Answers / Notes

3a. The person pulling the rope wins easily. A small effort applied to the rope at (c) moves a large resistance applied to the broom handle at (a) and (b).

3b. The person pulling the broom handle moves through the shortest distance. The end of the rope moves 6 times further than the broom handle.

4. This rope system works just like the super pulley of task 12. The broom corresponds to the movable wire pulley, and the stationary post corresponds to the fixed wire pulley. The number of rope strands connecting the two determines how far the end of the rope moves in relation to the broom handle: 6 support strands move the end of the rope 6 times farther.

Materials

☐ A strong smooth post fixed upright. A flagpole or tetherball post is appropriate.
☐ A strong cord or rope.
☐ A broom handle or piece of pipe.

(TO) build a working model of a wheel and axle. To calculate its ideal mechanical advantage.

WHEEL AND AXLE ○ Machines ( )

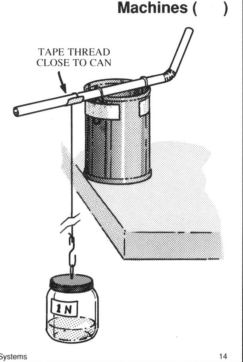

1. Build a working model of a wheel and axle machine from a tin can, paper clips, flexible straw and tape. Pull out the "arms" of the paper clips just a little, so the straw can turn freely.

2. The ideal mechanical advantage (IMA) of a machine is the ratio of *effort movement* to *resistance movement*.

$$IMA = \frac{\text{distance effort moves}}{\text{distance resistance moves}}$$

a. Calculate the IMA of your wheel and axle.
b. How can you increase the IMA on your wheel and axle machine?.

© 1989 by TOPS Learning Systems

Introduction
Draw a circle of any size on your blackboard, labeling its circumference and diameter. Use string to show that 3 circle diameters plus 1/7 more just fit around its circumference. No matter what size circle you draw, it is always true that the diameter divides into the circumference about 3.14 times. This leads to the relationship, C/D = 3.14 or C = πD.

Answers / Notes
1. *Placing the wheel and axle near the edge of the table will allow students to raise loads from the floor to the table. Caution students to cushion the glass jar with a coat or other material for soft landings, just in case there is an accidental free fall. Keep the thread close in, near the can, to avoid bending the straw. Hold the can to keep it from tipping, or add rocks for ballast.*

2a. The IMA for this soda straw wheel and axle is quite large. While the effort is applied along the circumference of a relatively large circle defined by the end of the soda straw (let's call this C), the resistance is only raised along the relatively small circumference of the soda straw axle (c). Here is a sample calculation for one particular brand of flexible straw:

$$IMA = \frac{\text{Distance Effort Moves}}{\text{Distance Resistance Moves}} = \frac{C}{c} = \frac{\pi D}{\pi d} = \frac{D}{d} = \frac{6.0 \text{ cm}}{0.6 \text{ cm}} = 10$$

2b. You can increase the IMA by increasing the diameter of the wheel (lengthening the straw), or by decreasing the diameter of the axle (using a narrower straw).

The actual mechanical advantage (AMA) of a machine is the ratio of the resistance force (its load) to the effort force. Because this takes frictional energy losses into account, the AMA is at best equal to, but usually less than, the IMA.

$$AMA = \frac{\text{resistance force}}{\text{effort force}}$$

Materials
☐ A medium-sized tin can, paper clips and masking tape.
☐ A flexible straw.
☐ The baby-food-jar resistance.
☐ Thread.

(TO) understand how gears work. To calculate rotational relationships between them.

SPIN YOUR WHEELS!   O   Machines ( )

1. Cut apart several rubber bands. Tie them together so they encircle two lids like this *without* stretching.

2. Stick paper-punch "washers" on 2 straight pins. Use cardboard as a pin cushion so you don't stick your fingers.

3. Fix each lid to cardboard, keeping them far enough apart to stretch the rubber band out just a little.

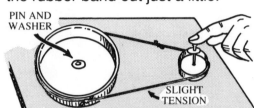

4. Bend over the pins underneath and secure with tape. Mark the rim of each lid with tape as well.

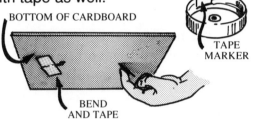

5. Spin your wheels. Ask yourself good questions, then answer them in a report.

© 1989 by TOPS Learning Systems

Answers / Notes

5. Here is an open-ended invitation for creative play and hands-on learning. Allow students the freedom to experiment in their own way, at their own pace. Most will begin by spinning wheels and counting rotations. Some may progress to spinning different wheels (the ones inside their own heads) as they learn to apply arithmetic in a predictive way.

To foster a mathematical approach make sure that string and rulers are available for those who want to measure. A small lid from a cooking oil bottle (C = 8.8 cm) and a larger lid from a mayonnaise jar (C = 22.0 cm), yield results like these:

If the small lid turns clockwise at 10 rpm, the large lid turns…

$$10 \text{ rpm} \times \frac{8.8 \text{ cm}}{22.0 \text{ cm}} = 4 \text{ rpm clockwise.}$$

If the large lid turns counterclockwise at 10 rpm, the small lid turns…

$$10 \text{ rpm} \times \frac{22.0 \text{ cm}}{8.8 \text{ cm}} = 25 \text{ rpm clockwise.}$$

Give the rubber band a half twist to make the wheels turn in opposite directions.

Materials

☐ Thin rubber bands, the longer the better. Very long rubber bands may not require cutting and joining in the first step. Smaller ones will.
☐ Scissors.
☐ A small and large lid. Punch small holes through the centers in advance with a small nail or thumbtack, using a hammer. The sides of each lid should be concave, allowing a rubber band to track around each circumference without slipping off.
☐ A paper punch.
☐ Straight pins.
☐ A flat piece of cardboard cut from a box.
☐ Masking tape.
☐ A ruler and string.

(TO) calculate how much horsepower can be generated by running up a flight of stairs.

HORSEPOWER Machines ( )

As machines replaced horses, people naturally estimated the power of a machine by how many horses it could replace. Today's engines are still rated in terms of *horsepower*.

Can you do the work of one horse? Use the equation to calculate how much horsepower you generate by running up a flight of stairs.

$$\text{Horsepower} = \frac{w \times h}{550 \times t}$$

w = your weight in pounds
h = height of stairs in feet
t = time in seconds

© 1989 by TOPS Learning Systems

Answers / Notes

This activity will generate a lot of enthusiasm, especially if you turn it into a class contest. Students will discover it is very difficult to generate a burst of power equivalent to one horsepower, even for just a few seconds.

A stopwatch is required for accurate measurements. The timer should stand at the top of the stairs, starting students from a resting position at the bottom. Assign a handicapped student or other reluctant stair-climber to this task, or do it yourself. Flying starts are illegal as well as dangerous. Supervise students closely to avoid injury and keep the noise level within a tolerable range.

Students may wish to run 2 or 3 trials, then select their best time. The fastest climber will not necessarily generate the greatest horsepower, since body weight must also be factored in. In fairness to all, you may wish to place a bathroom scale at the top of the stairs, so students can all weigh in on the same instrument.

After you gather all necessary data, return to the classroom for calculating horsepower and discussing the concepts of energy, work and power. Use hand calculators to minimize the time spent doing arithmetic. Ban them from your classroom if you want your students to practice challenging long division. In your follow-up discussion, consider these questions:

• *What happens to the work that you use to climb the stairs? Where does it go? (It increases your potential energy.)*
• *Do you do more work running up the stairs in a hurry? (No. Work does not depend on time.)*
• *Do you put out more power running up the stairs in a hurry? (Yes. Power measures the rate of doing work. Power = work/time.)*
• *Could you reduce the amount of work required to climb the stairs if you made them less steep? (No. You would reduce the force required to lift your body, but this would increase the distance you needed to run to achieve the same height. The product, force x distance, remains constant, independent of the slope of the stairs.)*

Materials

☐ A flight of stairs.
☐ A yard stick. A foot ruler and string will also serve.
☐ A stopwatch.
☐ A bathroom scale that measures in pounds.
☐ A calculator. (optional)

REPRODUCIBLE STUDENT TASK CARDS

☞ As you distribute these duplicated worksheets, **please observe our copyright notice** at the front of this module. We allow you, the purchaser, to make as many copies as you need, but forbid supplying photocopied materials to other teachers for use in other schools.

☞ TOPS is a small, not-for-profit educational corporation, dedicated to making great science accessible to students everywhere. Our only income is from the sale of these inexpensive modules. If you would like to help spread the word that TOPS *is* tops, please request multiple copies of our free **TOPS Ideas** catalog to pass on to other educators or student teachers. These offer a variety of sample lessons, plus an order form for your colleagues to purchase their own TOPS modules. Thanks!

Task Cards Options

Here are 3 management options to consider before you photocopy:

1. Consumable Worksheets: Copy 1 complete set of task card pages. Cut out each card and fix it to a separate sheet of boldly lined paper. Duplicate a class set of each worksheet master you have made, 1 per student. Direct students to follow the task card instructions at the top of each page, then respond to questions in the lined space underneath.

2. Nonconsumable Reference Booklets: Copy and collate the 2-up task card pages in sequence. Make perhaps half as many sets as the students who will use them. Staple each set in the upper left corner, both front and back to prevent the outside pages from working loose. Tell students that these task card booklets are for reference only. They should use them as they would any textbook, responding to questions on their own papers, returning them unmarked and in good shape at the end of the module.

3. Nonconsumable Task Cards: Copy several sets of task card pages. Laminate them, if you wish, for extra durability, then cut out each card to display in your room. You might pin cards to bulletin boards; or punch out the holes and hang them from wall hooks (you can fashion hooks from paper clips and tape these to the wall); or fix cards to cereal boxes with paper fasteners, 4 to a box; or keep cards on designated reference tables. The important thing is to provide enough task card reference points about your classroom to avoid a jam of too many students at any one location. Two or 3 task card sets should accommodate everyone, since different students will use different cards at different times.

LEVERS (1) ◯　　　Machines ( )

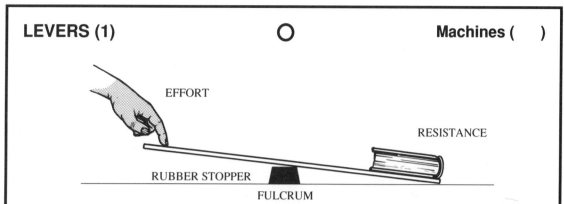

1. Lift a book with a lever: use a meter stick that pivots on a rubber stopper.
   a. Tell how to lift the resistance (the book) using a smaller effort (a smaller push from your hand). Draw a diagram.
   b. Tell how to lift the same resistance using a larger effort. Draw a diagram.

2. Did you apply the smaller effort through the same distance as the larger effort to lift the book? Explain.

© 1989 by TOPS Learning Systems

WORK ◯　　　Machines ( )

1. To do "work" in the scientific sense, you must apply a *force* through a *distance*. In each of the following activities decide if you are doing work. Explain your reasoning.
   a. Lift a book off the table.
   b. Hold a book perfectly still at arm's length for about 1 minute.
   c. Push a book across the table.
   d. Push against a wall as hard as you can.

WORK = Force x Distance

2. Call the force necessary to lift 1 book a "bk." Call the distance of your extended hand 1 "span."
   a. How much work is required to lift 2 books up 3 spans?
   b. How much work is required to lift 6 books up 1 span?
   c. Which requires more work, lifting 3 books up 3 spans or 2 books up 4 spans? Explain.

© 1989 by TOPS Learning Systems

LEVERS (2)   Machines (   )

1. Use masking tape to fix a rubber stopper to the 50 cm center of a meter stick and label it "C". Tape and label the left 10 cm mark "L" and the right 70 cm mark "R."

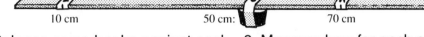

2. Balance some books against each other at points L and R. What can you discover?

3. Measure how far each point (L and R) move up and down. What can you discover?

4. If you push down with a force of 3 books at L…
    a. How many books can you just begin to lift up at R?
    b. Does this lever reduce your effort? Explain.

5. Suppose you apply this 3 book force down through a distance of .2 hand spans… a. Calculate the work you put *in* at L and the work you get *out* at R.
    b. Did the lever reduce your work? Explain.

© 1989 by TOPS Learning Systems

PAPER CLIP PULLEY (1)   Machines (   )

1. Get a baby food jar with a tight fitting lid. Fix a loop of thread to the exact center of the lid. Place 3 pieces of masking tape like this:

2. Get a spring scale. Add water to the jar so it weighs 1 Newton (100 grams).

3. Leak-proof the jar with plastic wrap. Trim excess, and label.

4. Record the minimum effort needed to HOLD your resistance in each system. Write your conclusions about (a) and (b); about (c) and (d).

A. NO PULLEY (right-side-up)

B. NO PULLEY (up-side-down) — USE PAPER CLIP HOOK

C. FIXED PULLEY (right-side-up) — PAPER CLIP PULLEY

D. MOVEABLE PULLEY (right-side-up) — PAPER CLIP PULLEY

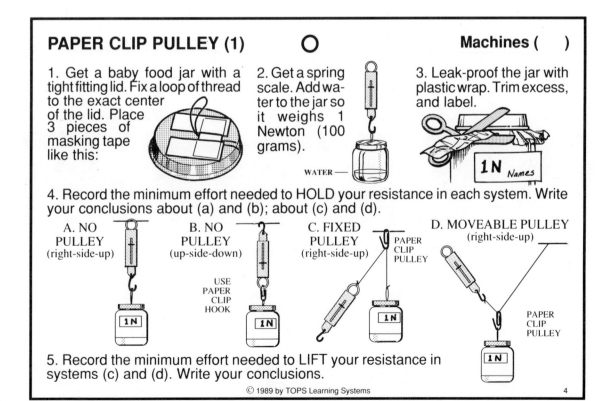

5. Record the minimum effort needed to LIFT your resistance in systems (c) and (d). Write your conclusions.

© 1989 by TOPS Learning Systems

PAPER CLIP PULLEY (2)   Machines ( )

1. How much work do you put *in* the moveable paper clip pulley when you pull the string a distance of 2 hand spans?

2. How much work does this pulley give *out* as it lifts the resistance?

3. Calculate the *efficiency* of the pulley.

$$\% \text{ Efficiency} = \frac{\text{Work}_{out}}{\text{Work}_{in}} \times 100$$

4. Does the pulley give...
   a. an effort advantage? Explain.
   b. a work advantage? Explain.

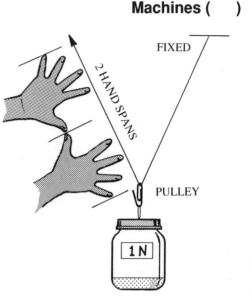

© 1989 by TOPS Learning Systems

WHEEL PULLEY   Machines ( )

1. Attach your load to a moveable wheel pulley.

2. Use your spring scale to calculate the efficiency of this pulley.

3. Why isn't the wheel pulley 100% efficient?

© 1989 by TOPS Learning Systems

COMBINATION PULLEY   Machines ( )

1. Twist two paper clips apart as shown.

2. Build a double paper clip pulley.

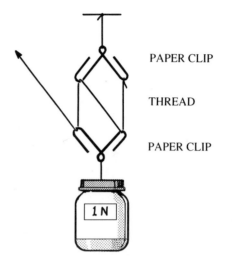

3. Calculate the efficiency of this machine. Show your work.

4. Suppose this pulley were friction free so that $Work_{in} = Work_{out}$. What effort would be needed to lift your load? Explain your reasoning.

© 1989 by TOPS Learning Systems

INCLINED PLANE (1)   Machines ( )

1. Fix a loop of string to the bottom of your load.

2. "Hitch" the scale to your "wagon" with paper clips and a piece of straw like this.

3. Cut and fold the paper protractor. Tape it and a straight piece of cardboard to your table so you can lift it through various angles.

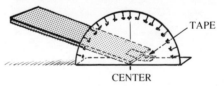

4. Measure the effort needed to pull your wagon up this incline while a friend holds it at each 10° of slope. Record you data in a table.

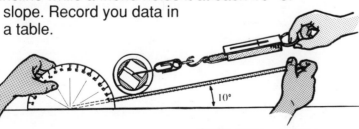

5. Plot a graph and interpret your results.

© 1989 by TOPS Learning Systems

INCLINED PLANE (2) Machines (   )

1. Divide two 4x6 index cards exactly in half the long way. Make a T-square from 3 pieces, and a 3-unit ruler from the last piece.

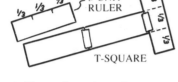

2. Use your tools to draw these 4 right triangles on scratch paper, then cut them out.

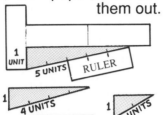

3. Use your protractor to fill in the value of all angles…

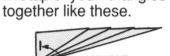

…Staple your triangles together like these.

4. Turn the triangles over to form 4 inclines for angling your cardboard ramp.

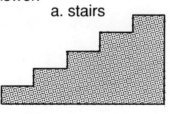

a. How much force must you use to lift your load from level A through 1 unit to level B? How much work does this require?

b. Assume this work is the same for each incline. Calculate the force you need to move your load from points C up 2, 3, 4, and 5 units to level B.

c. Check your answer on the 30° ramp.

© 1989 by TOPS Learning Systems

WHAT KIND OF MACHINE? Machines (   )

1. Classify each machine as a *lever* or *inclined plane*. Give reasons for your answer.

a. stairs

b. scissors

c. bolt

d. can opener

e. door knob

f. hatchet

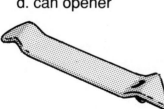
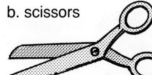
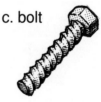

© 1989 by TOPS Learning Systems

WHAT KIND OF LEVER? ○   Machines ( )

1. Build each class of lever using a meter stick, a book and a rubber stopper. Tell if the machine reduces *force*, or *distance*, or *both*.

a. 1st Class   **b. 2nd Class**   **c. 3rd Class**

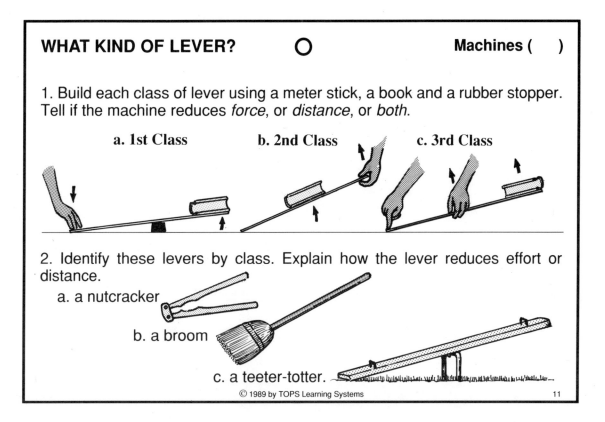

2. Identify these levers by class. Explain how the lever reduces effort or distance.
   a. a nutcracker
   b. a broom
   c. a teeter-totter.

SUPER PULLEY ○   Machines ( )

1. Build a super pulley from strong flexible wire. Attach the fixed pulley to a metal ring stand.

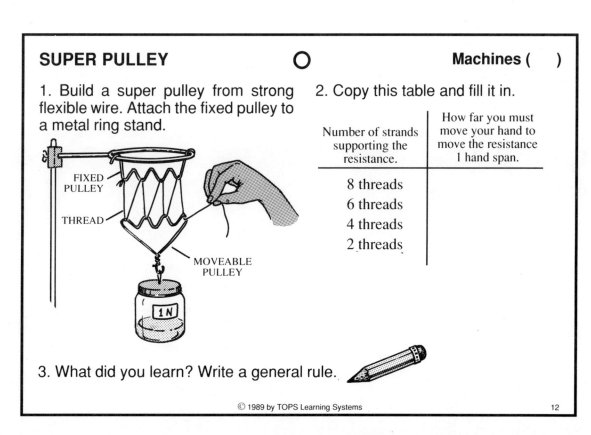

2. Copy this table and fill it in.

Number of strands supporting the resistance.	How far you must move your hand to move the resistance 1 hand span.
8 threads	
6 threads	
4 threads	
2 threads	

3. What did you learn? Write a general rule.

TUG-OF-WAR　　○　　　　　　　　　　　Machines (　)

1. Locate a strong, smooth post. You may need to go outside to find one.

2. Wind a rope around the post and a broom handle as illustrated.

3. Have a tug-of-war. You grip the broom at (a) and (b). Have a friend pull the rope at (c).
    a. Who wins the tug-of-war?
    b. Who moves through the shortest distance?

4. Compare this rope system to the super pulley you made before.

© 1989 by TOPS Learning Systems

WHEEL AND AXLE　　○　　　　　　　　Machines (　)

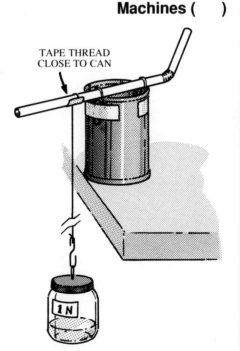

1. Build a working model of a wheel and axle machine from a tin can, paper clips, flexible straw and tape. Pull out the "arms" of the paper clips just a little, so the straw can turn freely.

2. The ideal mechanical advantage (IMA) of a machine is the ratio of *effort movement* to *resistance movement*.

$$IMA = \frac{\text{distance effort moves}}{\text{distance resistance moves}}$$

   a. Calculate the IMA of your wheel and axle.
   b. How can you increase the IMA on your wheel and axle machine?.

© 1989 by TOPS Learning Systems

SPIN YOUR WHEELS!   Machines ( )

1. Cut apart several rubber bands. Tie them together so they encircle two lids like this *without* stretching.

2. Stick paper-punch "washers" on 2 straight pins. Use cardboard as a pin cushion so you don't stick your fingers.

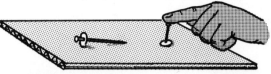

3. Fix each lid to cardboard, keeping them far enough apart to stretch the rubber band out just a little.

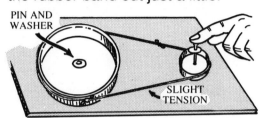

4. Bend over the pins underneath and secure with tape. Mark the rim of each lid with tape as well.

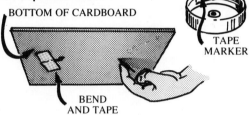

5. Spin your wheels. Ask yourself good questions, then answer them in a report.

© 1989 by TOPS Learning Systems   15

HORSEPOWER   Machines ( )

As machines replaced horses, people naturally estimated the power of a machine by how many horses it could replace. Today's engines are still rated in terms of *horsepower*.

Can you do the work of one horse? Use the equation to calculate how much horsepower you generate by running up a flight of stairs.

$$\text{Horsepower} = \frac{w \times h}{550 \times t}$$

w = your weight in pounds
h = height of stairs in feet
t = time in seconds

© 1989 by TOPS Learning Systems   16

PROTRACTOR

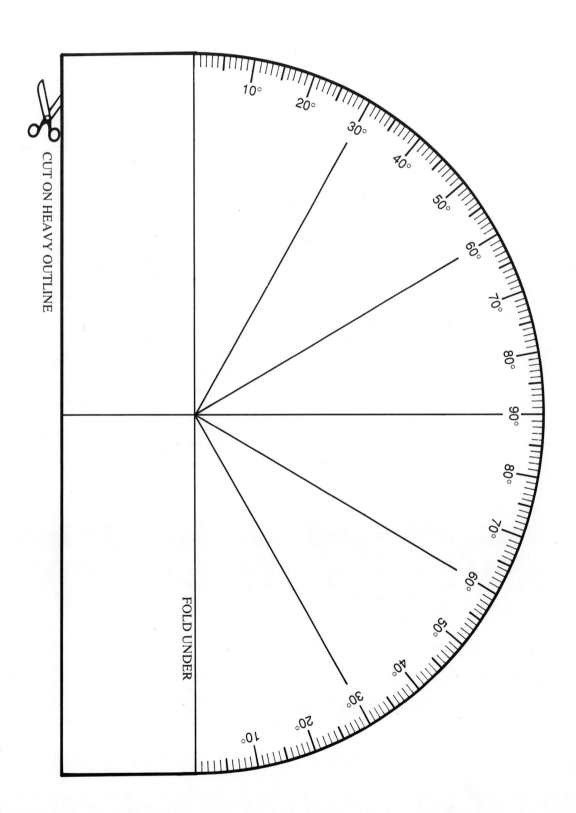

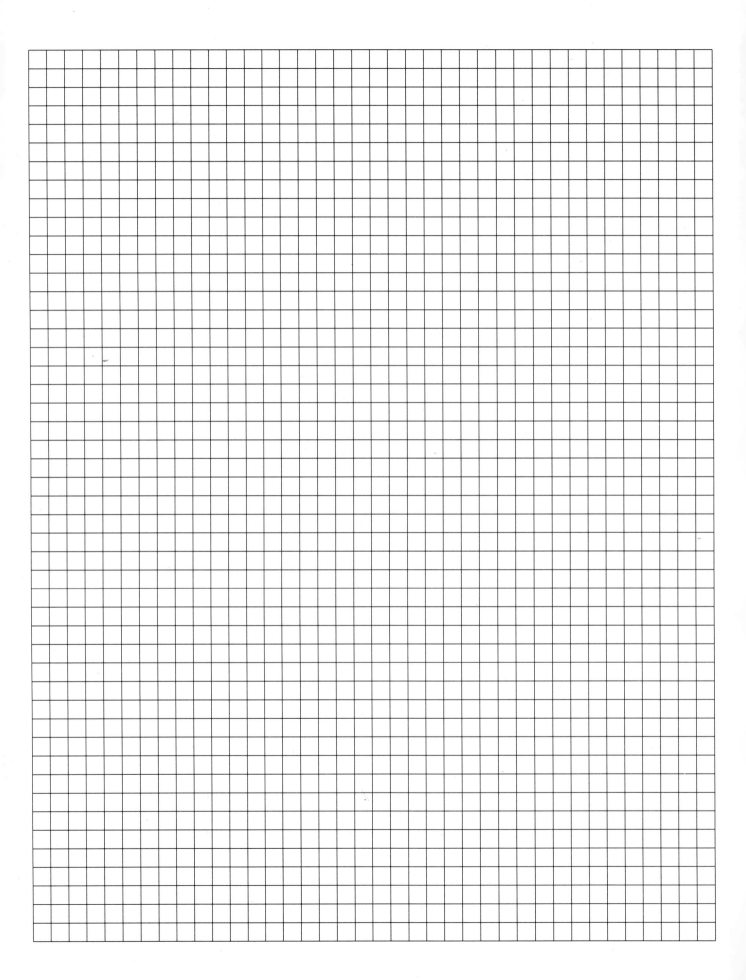